又高又凉快

High and Cool

Gunter Pauli

冈特·鲍利 著

王菁菁 译

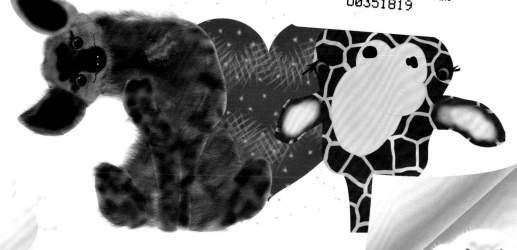

丛书编委会

主　任：贾　峰
副主任：何家振　郑立明
委　员：牛玲娟　李原原　李曙东　吴建民　彭　勇
　　　　冯　缨　靳增江

丛书出版委员会

主　任：段学俭
副主任：匡志强　张　蓉
成　员：叶　刚　李晓梅　魏　来　徐雅清　田振军
　　　　蔡雩奇

特别感谢以下热心人士对译稿润色工作的支持：

姜竹青　韩　笑　杨　爽　周依奇　于　哲　阳平坚
李雪红　汪　楠　单　威　查振旺　李海红　姚爱静
朱　国　彭　江　于洪英　隋淑光　严　岷

目录

又高又凉快	4
你知道吗？	22
想一想	26
自己动手！	27
学科知识	28
情感智慧	29
艺术	29
思维拓展	30
动手能力	30
故事灵感来自	31

Contents

High and Cool	4
Did you know?	22
Think about it	26
Do it yourself!	27
Academic Knowledge	28
Emotional Intelligence	29
The Arts	29
Systems: Making the Connections	30
Capacity to Implement	30
This fable is inspired by	31

一只长颈鹿想喝水。看着他使劲低下头才能够到水的笨拙样子，一群鬣狗歇斯底里地大笑起来。

"你们笑什么？"长颈鹿冷静地问。

A giraffe is trying to have a drink. It does not help that a couple of hyenas are laughing hysterically at his clumsy attempts to get down low enough to reach the water.

"What are you guys laughing at?" asks the giraffe, coolly.

一只长颈鹿想喝水

A giraffe is trying to have a drink

你看起来真好笑！

You look so funny!

"看你扭动着关节，伸出长脖子的样子，看起来真好笑。你一定经常脖子疼吧。"鬣狗首领咯咯笑着说。
"你们才是唯一让我脖子疼的家伙！我有强壮的颈部肌肉——我可以很容易地通过摇摆脖子击倒敌人。而且我的心脏是个超级强大的血泵。"

"You look so funny when you wiggle on your knees with your long neck stretched out. You must often have neck pain," giggles the leader of the hyena pack.
You are the only pains I have in my neck! I have strong neck muscles – I can easily knock out my enemies by swinging my neck at them. And my heart is a super-strong pump."

"你可骗不了我。"

鬣狗说,"世界上最强大的心脏是鲸的心脏。他号称'心脏之王',这可不是徒有虚名。"

"他当然是心脏之王,不过他生活在海洋里。"长颈鹿回答道,"由于我的头比我的心脏高两米多,我需要巨大的压强和非常强烈的心跳才能把足够的氧气输送到我的大脑。"

"You can't fool me," says the hyena. "The greatest pump on earth is the whale's heart. He is not known as the King of Hearts for nothing."

"Of course he is the king, but he lives in the ocean," answers the giraffe. "As my head is more than two metres above my heart, I need huge pressure and a very strong heartbeat to get enough oxygen to my brain."

……世界上最强大的心脏是鲸的心脏

...greatest pump on earth is the whale's heart

我有身上的斑块来降温!

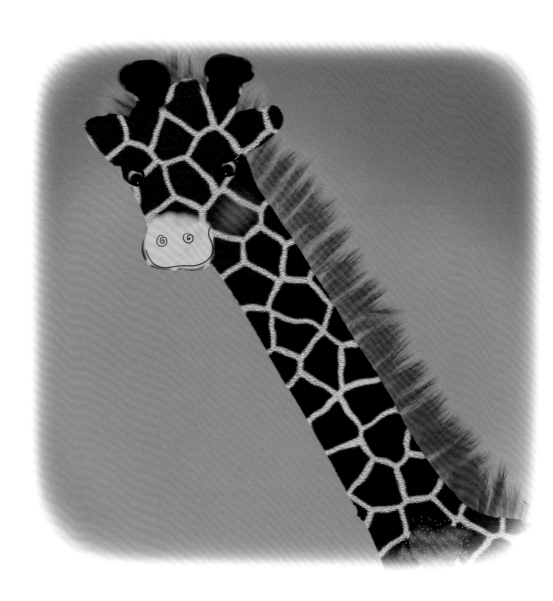

I also have my patches to keep me cool!

"我的静脉和动脉会因为这种巨大的压强而爆裂的!我们鬣狗更喜欢大一些的心脏和缓慢一些的心跳。"

"我的静脉壁比你们鬣狗的皮肤还要厚。幸好我有身上的斑块来降温!"

"你那些深色的斑块一定让你的身体因为太暖和而不舒服。"

"My veins and arteries would pop with that kind of pressure! We hyenas prefer a bigger heart and a slower beat."

"The walls of my veins are thicker than your hyena skin. Fortunately I also have my patches to keep me cool!"

"Your dark patches must make your body too warm for comfort."

"啊，我知道你们了解斑马，"长颈鹿说，"我们的朋友斑马用黑白相间的外衣来调节体温！而我皮肤上的深色斑块布满血管，这是我用来释放体热的地方。"

"这听起来像是一扇释放热气的窗户。"鬣狗说。

"的确如此，这就是我用来保持身体凉爽的方式。"

"Ah, I see you know about the zebra," says the giraffe. "Our friend the zebra uses his black-and-white coat to create his own air conditioning! My dark patches on my skin are full of blood vessels, and that's where I release body heat."

"That sounds like a window letting out the heat," says the hyena.

"Absolutely, and that is how I keep my body cool."

我们的朋友斑马……

Our friend the zebra...

你是如何使你的头部保持凉爽的呢?

How do you keep your head cool?

"但是你的头部总是在太阳下晒着,从金合欢树上采摘叶子吃。你是如何使你的头部保持凉爽的呢?"

"所有这些斑块,特别是脖子上的那些,可以像空调一样工作,这样即使我在太阳下,我头部的温度也总是能比我身体的温度低十度。"

"But your head is always in the sun, plucking leaves from the acacia tree. How do you keep your head cool?"

"All these patches, especially the ones on my neck, work like an air conditioner so that even though I am in the sun, the temperature of my head is always ten degrees lower than my body."

"真的吗？"鬣狗用新的眼光重新审视长颈鹿的外衣。

"是的，血液通过心脏被压送到身体上方，让我的头部保持凉爽。"

"大自然不需要电池就可以做到这些，这不是太神奇了吗？"鬣狗惊讶地说。

"Really?" The hyena looks at the giraffe's coat with new eyes.

"Yes, my body is designed to pump blood upwards and keep my head cool."

"Isn't it simply amazing what nature can do without batteries?" muses the hyena.

大自然不需要电池就可以做到这些，太神奇了

Amazing what nature can do without batteries

我们是原始的"智能电网"

We are the original "smart grids"

"是的，大自然中的任何事物都不需要电池或燃烧火焰来制造能量。我们都是自力更生。我们是原始的'智能电网'。"

"天哪！现在我知道当我们这么多鬣狗围着你的时候你是如何保持如此冷静的头脑了。"鬣狗轻声地笑道。

"Yes, nothing in nature needs batteries, or burning fires, to create power. We make it all ourselves. We are the original 'smart grids'."

"Goodness! Now I know how you keep such a cool head with us hyenas around you," chuckles the hyena.

"是的，记住千万别惹我。我可以用强壮的腿将你狠狠踢倒，让你甚至你的子孙都会铭记在心！"

……这仅仅是开始！……

"Yes, and remember not to mess with me: with these strong legs I could kick you so hard even your children would remember it!"

...AND IT HAS ONLY JUST BEGUN!...

……这仅仅是开始!……

...AND IT HAS ONLY JUST BEGUN!...

Did You Know?
你知道吗？

Giraffes are the tallest mammals in the world. Even newborn calves are taller than most humans.

长颈鹿是世界上最高的哺乳动物。即使是新生的长颈鹿幼仔也比大多数人类要高。

Giraffes spend their lives standing upright. They even sleep and give birth standing up.

长颈鹿一生都保持着站立状态。他们甚至站着睡觉和生育后代。

Males fight by necking: hitting each other with their necks.

雄性长颈鹿用脖子打架，即用他们的脖子互相击打对方。

No two giraffes have the same pattern of spots. Their coats are as unique as snowflakes or fingerprints.

任何两只长颈鹿都不会拥有同样的斑块图案。他们的外衣具有像雪花或指纹一样的独特性。

The scientific name given to the giraffe, Giraffa camelopardalis, means a camel painted like a leopard.

长颈鹿的学名是 *Giraffa camelopardalis*，意思是画着豹纹的骆驼。

Spotted hyenas live in clans of up to 80 members and eat flesh, skin, bones and even animal droppings. Hyenas have their own latrines, called middens, where clan members leave their faeces.

黑斑鬣狗喜欢群居，群居成员多达 80 只，他们以肉、皮毛、骨头，甚至是动物的粪便为食。鬣狗有自己的厕所，叫做粪堆，族群成员都会在那里排下粪便。

𝒜 hyena's 'laugh' alerts other clan members to a good source of food. They have the strongest jaws of any land animal and are able to pulverise the entire skeleton of their prey.

鬣狗用"笑声"提醒其他族群成员找到了好的食物来源。他们拥有陆地动物中最强壮的下颌,能够将猎物的整个骨架咬碎。

𝒯he highest-ranking male hyena is subordinate to the lowest-ranking female. Hyenas form highly social groups and will snuggle together in a shaggy pile. When faced with a threat, a hyena may play dead.

最高级别的雄性鬣狗的地位低于最低级别的雌性鬣狗。鬣狗们构建高度社会化的群体,他们会蜷缩在一起形成毛茸茸的一堆。当遇到威胁时,鬣狗可能会装死。

Think About It
想一想

Did you ever think that an animal could laugh?

你认为动物真的会笑吗?

长颈鹿是如何通过长脖子来降温的?

How did it come about that giraffes grew tall necks to cool down?

How is it that everything in nature has power, and that no creature uses batteries?

自然界的一切事物都有自己的能量,而且任何生物都不需要电池,这是怎么回事?

如果有一群鬣狗围着你,你能保持冷静吗?

Would you be able to keep calm with a pack of hyenas circling around you?

Do It Yourself!
自己动手!

Let's play a clan game. If one person is considered the biggest and the strongest, how many others, working as a team, are needed to push him or her off balance? If someone is considered the smartest, how many team members are needed to find faster and better answers to the problems and challenges posed? A team that is tightly knit into a clan will always win, even against the smartest and the strongest.

让我们来玩一个集体游戏。如果有一个被认为是最高大最强壮的人,要想推倒他/她,需要一个由多少人组成的团队?如果有一个被认为是最聪明的人,要解决他/她提出的问题或挑战,需要多少队员才能更快更好地找到答案?即使是面对最聪明和最强壮的对手,一支紧密团结的队伍最终总会赢得胜利。

TEACHER AND PARENT GUIDE

学 科 知 识
Academic Knowledge

生物学	鬣狗共有四大家族：斑鬣狗、棕鬣狗、条纹鬣狗和土狼；鬣狗跟狗没有亲缘关系；鬣狗是领地性动物，用富含钙质的白色粪便来标明自己的领地；黑斑鬣狗拥有复杂的肢体语言，包括问候仪式；鬣狗的心脏比狮子的心脏大一倍；鬣狗可以奔跑几英里，时速能达到60公里；长颈鹿有7节颈椎，就像人类一样；长颈鹿在雨季以阔叶和落叶植物为食，其他季节以常绿植物为食。
化 学	高钙质的饮食导致鬣狗排出白垩样的粪便；通过产生气味的细菌制造的"社会气味"帮助鬣狗们交流。
物 理	长颈鹿的脖子越长，体表释放的热量就越多；在大自然中，能量的产生都不需要电池。
工程学	"智能电网"是一种使用信息和通信技术来管理电力生产和消费的电力网络；双层玻璃窗、甚至三层玻璃窗的设计是为了在冬天将热量保存在室内，防止冷空气进入。
经济学	鬣狗是一种关键物种，他们无法在某个区域存活下去就说明了当地生态系统的退化，这意味着经济会受到影响，一开始是旅游业，然后就是其他重复性的生产活动，比如畜牧业、野生水果和蔬菜的采摘。
伦理学	鬣狗一直被认为与巫术有关，这种说法甚至在现在仍然被媒体与娱乐业不断强化，即使科学上已有定论，但要扭转大众的普遍观点还是非常困难。
历 史	在欧洲的拉斯科洞窟中可以找到有关鬣狗的岩画；亚里士多德和老普林尼用文字记录了鬣狗的发现；在来自东非的跶巴娃神话中，鬣狗是第一个将太阳带到地球、给寒冷的地球带来温暖的动物；长颈鹿曾在罗马大竞技场展出；中国人第一次看到长颈鹿是在1414年，他们以为这是神话传说里被称为"麒麟"的有蹄动物。
地 理	长颈鹿是非洲的代表性动物，也是当地旅游经济的卖点。
数 学	长颈鹿的血压很高（280/180毫米汞柱），是人类血压的两倍，其心脏跳动次数高达每分钟170次，长颈鹿颈部的长度每增加15厘米，左心室的室壁就会增厚0.5厘米。
生活方式	鬣狗的一种生活方式是偷窃——他们可以偷走狮子口中的新鲜猎物；当遇到威胁的时候，鬣狗会装死，就像负鼠一样。
社会学	鬣狗喜欢群居生活，但是却被大多数西方文化和非洲文化认为是愚蠢、懦弱和丑陋的。
心理学	一旦我们对某个人(或某只动物)下了定论，我们就很难改变想法，即使有科学依据也依然很难。
系统论	鬣狗可以作为检验生态系统健康状况的指标之一。

教师与家长指南

情感智慧
Emotional Intelligence

长颈鹿

当鬣狗们嘲笑长颈鹿试图喝水的样子时,他很生气。然而,长颈鹿有自知之明,解释了他的身体从颈部到心脏是如何运行的。长颈鹿在他的表述中非常明确地证实他拥有陆地动物中最强大的心脏。长颈鹿还意识到其他物种存在的作用和重要性,特别是鲸,他还认识到自己在生态系统中的地位。长颈鹿包容了鬣狗的无礼之言,并准备好解释他的独特性:他有长长的脖子,上面的斑块可以帮助他保持凉爽。当鬣狗表现出感兴趣时,长颈鹿进一步解释了更多细节。长颈鹿对于自己有足够清晰的认识,将自己的能量、泵送和冷却系统形象地比喻成真正的"智能电网"。长颈鹿以对鬣狗的警告结束了对话,证实了他的自信心。

鬣狗

鬣狗对长颈鹿表现出的工程学奇迹没有表示出一点尊重,但是对于可能出现的头疼表示了关心。鬣狗非常自信,认为长颈鹿的话愚弄不了她。然而,鬣狗并没有专心听,因为她漏掉了长颈鹿清晰明确的说明。鬣狗意识到了自己的局限性,并对血泵的规格和心跳问题发表相关评论。鬣狗分享了自己的传统知识,虽然被回绝,但是很快用刚学到的新知识重新调整了自己的想法。鬣狗意识到自己掌握的原有信息与实际不符,因此表现出了对长颈鹿的尊重。

艺术
The Arts

众所周知,鬣狗会发出非常特别的声音。寻找一家声音图书馆,那里可以为你提供各种不同的大笑声和傻笑声。鬣狗还有非常明显的肢体语言。让我们试着表演鬣狗生活中的一天。学习鬣狗的身体姿态、动作和鬼脸。选择一组人专门发出声音,另一组人配合做出鬣狗的姿势和动作。让观众通过两组人发出的声音或做出的肢体表达来判断每一次表演的含义。

TEACHER AND PARENT GUIDE

思维拓展
Systems: Making the Connections

生活中总是有令人惊奇的发现。鬣狗已经演变成非洲大草原上最重要的捕食者，是关键物种之一。借助其智力和毅力，鬣狗能够度过大多数艰难时刻。由于鬣狗位于食物链或者说生物金字塔的顶端，可以作为对生态系统质量的评判标准。如果鬣狗濒临灭绝，就说明在各个营养级的所有生物都受到威胁。这也就是鬣狗族群的健康状况和多样性能为生态系统的状况提供保障的原因。鬣狗的消化系统将钙质以阳离子形式返回到土壤，成为一种非常适合帮助植物生长的元素。泥土中结合入钙质，可以产生更好的土壤孔隙度和土壤透气性，以保障排水和根部生长。长颈鹿也是一种同样重要的关键物种，身高5.5米，是地球上最高的动物，能够轻松够到树冠，而其他食草动物却无法做到。长颈鹿和他最喜爱的食物——金合欢树开展进化竞赛。随着长颈鹿长高，金合欢长出刺和丹宁。长颈鹿又进化出结实的嘴和长舌头来剥离树叶以免被刺伤。长颈鹿避免食用丹宁，而且金合欢只在长颈鹿开始觅食树叶几分钟之后才会产生这种毒素，这样长颈鹿就不会将树叶全部吃光。长颈鹿和鬣狗的故事提供了独特的示范，证明了进化是如何远远超越工程学和人性化设计的。

动手能力
Capacity to Implement

列出一份鬣狗和长颈鹿的特征清单。注意每个物种今天的外表都是与其他物种，而不仅仅是与同一物种相互竞争作用的结果。详细记录下使这些动物成为今天的样子的影响因素：仅仅通过观察这两种关键物种，生命的网络就变得非常清晰。现在，想一想有哪些物种需要依赖鬣狗和长颈鹿而生存。注意围绕鬣狗和长颈鹿的生命网络是如何分别慢慢融合到更大的生命网络中。每当你在思考一个项目或者创建一个新产业时，画出一份生命网络。这样会帮助你理解潜在的协同效应，让你事半功倍，帮助你更容易做出调整。